HELLO?

'LOR? HI, IT'S ME? YOU THINK YOUR MOM WILL LET YOU COME OVER?

SHE DOESN'T CARE. TOO BUSY WATCHING TV.

BRING ME SOMETHING TO EAT?

OKAY, SEE YA IN A BIT!

YOUR MOM FORGOT TO SHOP AGAIN?

YUP.

MOM, I'M GONNA GO TO ESTELLE'S, OKAY?

MOM? MOM, I'M GOING OUT.

YOU'RE GONNA FRY YOUR EYEBALLS.

SHUT UP! I CAN'T HEAR THE TV!

WHATEVER.

IN 1989, MEREDITH VIEIERA WON FOUR EMMY AWARDS--NO.

HER INVESTIGATIONS EARNED VIEIRA... VIEIRA'S WORK ON THE TV NEWS PROGRAM WEST 57TH...

FIVE MILES AND ONE HOUR LATER
---AND EARNED HER FIFTH EMMY FOR HER WARD 5A SEGMENT ON 60 MINUTES.
'ELLE, IT'S ME, BUZZ ME IN!
BRZZZZZZZZ!
THANKS FOR COMING.
OH, MY GOD! IF YOUR MOM SAW THAT OUTFIT, SHE WOULD FREAK OUT.
YEAH, WELL, SHE WOULD HAVE TO BE HOME TO SEE IT.
EVEN IF SHE WAS HOME, I DOUBT SHE WOULD NOTICE.
HA, HA, YEAH!
OH, SHE WOULD NOTICE. SHE WOULD SAY, "WHAT AN UNFORTUNATE COLLECTION OF CLOTHING YOU'VE ASSEMBLED, DAUGHTER."
WHAT'CHA BRING ME TO EAT? I'M STARRRRVING.
ALL KINDS OF STUFF.

MEREDITH BECAME AN INVESTIGATIVE REPORTER FOR WCBS-TV IN CHICAGO AND COVERED THE 1980 REPUBLICAN NATIONAL CONVENTION.

SHE WORKED AS A SUBSTITUTE TV NEWS ANCHOR...

...AND EARNED A FRONT PAGE AWARD FROM HER SERIES ON CHILD MOLESTATION.

IN 1982, MEREDITH TOOK A POSITION AT CBS NEWS IN CHICAGO. SHE BECAME A CORRESPONDENT IN 1984--

UM...

IT'S...A LITTLE DRY, ISN'T IT?

MEREDITH WASN'T AN OVERNIGHT SENSATION. SHE HAD TO WORK REALLY LONG AND HARD TO GET WHERE SHE IS TODAY.

GOOD POINT. CONTINUE.

I'LL SKIP AHEAD TO SOME JUICY STUFF... UM...

...WHILE AT CBS, MEREDITH MET THE MAN WHO CHANGE HER LIFE FOREVER.

OH, THAT'S GOOD. DRAMATIC.

YEAH, RIGHT?

DESPITE COHEN'S SARCASTIC REMARKS ABOUT HER MATURITY, MEREDITH SAID THERE WAS A SPARK BETWEEN THEM.

I THOUGHT "FOOL, JERK," WHATEVER... IT WASN'T THAT AN ARROW WENT THROUGH MY HEART IN THAT MOMENT, BUT I THOUGHT, "I'M GONNA MARRY THAT GUY."

ON THEIR SECOND DATE, RICHARD CONFIDED IN MEREDITH THAT HE SUFFERED FROM MULTIPLE SCLEROSIS.

TO RICHARD, HAVING M.S. WAS...

... JUST SOMETHING TO ADDRESS...I JUST WANTED HER TO HAVE THE BENEFIT OF THAT INFORMATION.

ESTELLE, HONEY, YOU KNOW HOW I FEEL ABOUT YOU HAVING COMPANY OVER WITHOUT MY PERMISSION.

MOM!

YOU SHOULD BE GETTING HOME, CHANCELOR.

...YES, 612 WHARF AVENUE, MH-HM...

WE WERE JUST WORKING ON OUR SCHOOL PROJECT, MAMA.

THERE'S A CAB DOWNSTAIRS. YOU CAN TALK TO ESTELLE TOMORROW AT SCHOOL.

THANK YOU, MS. SIMMONS.

G'NIGHT, 'ELLE!

HOW MANY TIMES MUST WE HAVE THIS CONVERSATION?

BUT--

ESTELLE LILLY, DON'T YOU TALK BACK TO ME.

I'M SORRY, MAMA. UM, CAN YOU COME TO WATCH ME TOMORROW?

FIVE MILES AND ONE HOUR LATER
UH-OH.
WHERE HAVE YOU BEEN? IT'S AFTER TEN.
I TOLD YOU I WAS GOING TO ESTELLE'S TO WORK ON MY REPORT.
NEVER TOLD ME NO SUCH THING.
YEAH, YOU DIDN'T HEAR ME BECAUSE YOU WERE PASSED OUT.
CHANCE! I DID NOT *PASS OUT* I FELL ASLEEP.
YOUR DADDY, GOD REST HIS SOUL, WAS THE DRUNK NOT ME.

HEY, I LOOKED FOR YOU AT LUNCH.

SORRY, WE HAD A "CRISIS" AT THE SCHOOL NEWSPAPER.

NERVOUS?

NO, I JUST WISH...

I KNOW. MY MOM'S NOT HERE EITHER.

FOR OUR REPORT ON WOMEN ROLE MODELS...

...WE CHOSE MARY-KATE AND ASHLEY OLSEN.

OH, MY GOD, REALLY?

I KNOW, RIGHT?

BE NICE, GIRLS. GET READY. YOU'RE UP NEXT.

MEREDITH FOLLOWED UP HER MULTIPLE EMMY WINNING FOUR-YEAR RUN ON *WEST 57TH* BY BECOMING A CORRESPONDENT AND CO-EDITOR FOR *60 MINUTES*.

AT *60 MINUTES*, MEREDITH AGGRESSIVELY PURSUED SOME OF THE MOST IMPORTANT AND CONTROVERSIAL ISSUES OF THE DAY, EARNING RECOGNITION FOR HER *"WARD 5A"* STORY ON THE FIRST A.I.D.S. WARD IN SAN FRANCISCO...

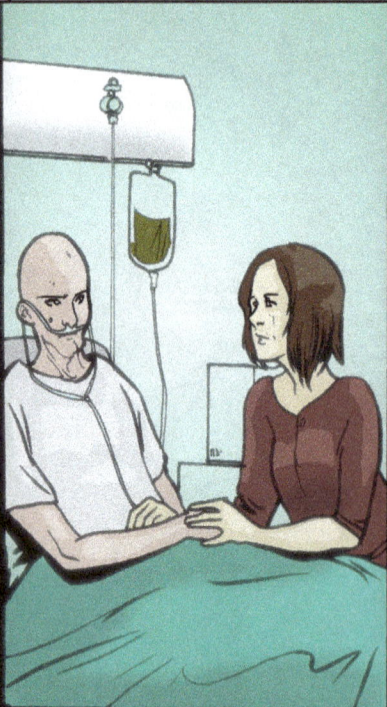

...AND A FIFTH EMMY FOR HER STORY *"THY BROTHER'S KEEPER"* ABOUT CHRISTIANS WHO RISKED EVERYTHING TO SAVE THEIR JEWISH BROTHERS AND SISTERS DURING THE HOLOCAUST.

DESPITE BEING A, UM, SUCCESSFUL TV JOURNALIST, MEREDITH, UM...MADE HER FAMILY A TOP PRIORITY.

MEREDITH WAS FORCED TO LEAVE 60 MINUTES WHEN THE EXECUTIVE PRODUCER REFUSED TO ALLOW HER TO CONTINUE TO WORK PART-TIME WHEN SHE BECAME PREGNANT WITH HER SECOND CHILD.

MEREDITH'S DECISION WAS CRITICIZED BY SOME, BUT SHE SAYS THAT SHE HAS NO REGRETS.

TO ME, HAVING IT ALL MEANS BEING ABLE TO SET THE STANDARDS THAT YOU WANT FOR YOUR LIFE.

AFTER A BRIEF STINT ON CBS' VERDICT, MEREDITH BECAME CO-ANCHOR OF THE CBS MORNING NEWS IN 1992.

PRESS

MEREDITH MOVED TO ABC IN 1993 AND SPENT FOUR YEARS AS CHIEF CORRESPONDENT ON *TURNING POINT*, EARNING ANOTHER EMMY FOR A STORY ON WHITE SUPREMACISTS.

WHITE NATION

THE CANCELLATION OF THE SHOW WAS A *"TURNING POINT"* FOR MEREDITH, AS WELL.

I WAS A REPORTER WHO DIDN'T WANT TO REPORT BECAUSE IT REQUIRED A TREMENDOUS AMOUNT OF TRAVEL... I HAD TO REINVENT MYSELF.

MEREDITH HELPED LAUNCH THE VIEW IN 1997, ACTING AS CO-HOST AND MODERATOR UNTIL 2006.

500

TAKE SOME TIME TO ENJOY THE VIEW.

IN THAT TIME, SHE ALSO RAISED THREE CHILDREN AND HELPED HER HUSBAND THROUGH TWO BOUTS OF COLON CANCER.

MEREDITH LEFT *THE VIEW* IN 2006 TO REPLACE KATIE COURIC ON NBC'S *TODAY SHOW*, WHICH SHE STILL CO-HOSTS WITH MATT LAUER.

SHE, UM...*DATELINE NBC*... UM, BEEN THE HOST OF *WHO WANTS TO BE A MILLIONAIRE* SINCE 2002...

...UH, SHE...SOME HAVE CRITICIZED HER BECAUSE OF HER RISQUÉ SENSE OF HUMOR AND FASHION, BUT SHE DOESN'T LET IT FLUSTER HER.

I'M A JOURNALIST, YES, BUT I'M ALSO A WOMAN, AND I'M NOT ASHAMED OF BEING A WOMAN.

DESPITE JUGGLING WORK AND FAMILY, MEREDITH SOMEHOW FINDS THE TIME TO PURSUE HER OTHER PASSIONS, LIKE ACTING...

...AND CHARITY WORK. SHE HAS BEEN AWARDED THE SAFE HORIZON CHAMPION AWARD, A CITY OF HOPE WOMAN OF THE YEAR AWARD, AND THE MOTHER OF THE YEAR AWARD FROM THE PAJAMA PROGRAM.

MEREDITH HAS ALSO BEEN HONORED BY THE ANTIDEFAMATION LEAGUE.

THOUGH HER INCREDIBLE TALENTS HAVE BROUGHT HER WEALTH, SUCCESS, AND FAME, SHE CONSIDERS ALL OF IT SECONDARY TO HER FAMILY...

I JUST WANT TO DO MY JOB AND GO HOME. I KNOW THAT'S WHAT MY FAMILY EXPECTS.

...WHICH IS WHAT WE HAVE FOUND TO BE THE MOST INSPIRING QUALITY OF OUR WOMAN ROLE MODEL, MEREDITH VIEIRA.

HOW'D WE DO, MS. DE PAULA?

ARE WE GETTING A'S?

OH, MY GIRLS! THAT WAS WONDERFUL. LISTEN TO THAT APPLAUSE.

CLAP CLAP CLAP CLAP CLAP

BLUEWATER COMICS

★ FEMALE ★ FORCE ★

Meredith Vieira

Brent Sprecher ———————————— Writer

Alex Lopez ———————————————— Penciler

Willie Jimenez ———————————— Colorist

Wilson Ramos Jr. ———————————— Letterer

Darren G. Davis ———————————— Graphics

Darren G. Davis
Publisher

Jason Schultz
Vice President

Aha Maree
Creative Development

Crystal VanDiver
Director

Vinnie Tartamella
Cover

Chad Jones
Production

BLUEWATER COMICS

Janda Tithia
Coordinator

Scott Davis
Media Manager

Kim Sherman
Marketing Director

Vonnie Harris
New Business

Adam Ellis
Coordinator

www.bluewaterprod.com

With your donated dollars and volunteer hours, we work tirelessly to erase hate from every corner of America through our programs.

SPEAKING ENGAGEMENTS

Since Matt's death in 1998, Judy and Dennis have been determined to prevent others from similar tragedies. By sharing their story, they are able to carry on Matt's legacy.

HATE CRIMES REPORTING

Our work to improve reporting includes conducting trainings for law enforcement agencies, building relationships between community leaders and law enforcement, and developing policy reform in reporting practices.

LARAMIE PROJECT

MSF offers support to productions of The Laramie Project, which depicts the events leading up to and after Matt's murder. It remains one of the most performed plays in America.

MATTHEW'S PLACE

MatthewsPlace.com is a blog designed to provide young LGBTQ+ people with an outlet for their voices. From finance to health to love and dating, and everything in between, our writers contribute excellent material.

Erase Hate

Matthew
Shepard
Foundation
embracing diversity